SO-BRF-264

CHILDREN'S SCHOOL OF SCIENCE

STRANGE MONSTERS OF THE SEA

by Richard Armour

Paul Galdone
drew the pictures

CHILDREN'S SCHOOL OF SCIENCE

McGRAW-HILL BOOK COMPANY
New York   St. Louis   San Francisco

ACKNOWLEDGMENTS

This book is playful, but I hope it is also correct in its facts about the strange creatures described. In addition to reading numerous books in marine biology, I had the help of two biologists who checked everything carefully. These two scientists, to whom I am greatly indebted, are Dr. Larry Oglesby, Professor of Biology at Pomona College, and Dr. Robert Feldmeth, Associate Professor of Biology in the joint science department of Claremont Men's College, Pitzer College, and Scripps College. I am grateful to them for suggesting books, answering numerous questions, and finally reading the entire manuscript. I take the responsibility, however, for any faults that may remain.

R.A.

Library of Congress Cataloging in Publication Data

Armour, Richard Willard
Strange monsters of the sea.

SUMMARY: Describes the characteristics of such unusual sea creatures as the giant grouper, deep-sea spider, zebrafish, and dragonet.

1. Sea monsters—Juvenile literature. [1. Sea monsters] I. Galdone, Paul. II. Title.
QL89.2.S4A75   591.92   78-11263
ISBN 0-07-002294-1

Text copyright © 1979 by Richard Armour. Illustrations copyright © 1979 by Paul Galdone. All Rights Reserved. Printed in the United States of America. No part of this publication may be reproduced, stored in a retrieval system, or transmitted, in any form or by any means, electronic, mechanical, photocopying, recording, or otherwise, without the prior written permission of the publisher.

123456789   RABP   7832109

Far more of our world is sea than land,
And it's hard for people to understand,
Since they live on land, just how it must be
To live, all their lives,
In the depths of the sea.

Yet creatures there are that live down there,
Where they can't see the clouds
And can't breathe the air,
Where no laughter is heard
And no books are read,
And keeping alive's what they do instead.

Well, here are a few (there are many more)
That live in the ooze of the ocean's floor
Or swim or wriggle
Their very own way
In search of their unsuspecting prey.

DEEP-SEA ANGLERFISH

A fisher of fish is the Anglerfish,
Equipped for its task in all ways it could wish.
It's shaped like a pear,
And as wide as it's high.
You'd think it quite weird if you happened by.
But I'm sorry to tell you,
And hope you won't frown,
That this is unlikely—
It lives two miles down!

The Angler possesses a rod and a line
Sticking up from its head
That for fishing is fine.
At the end of the line is what looks just like bait,
With a light on it, too,
Luring fish to their fate.

Yes, angle for fish
Does the Angler with zeal,
Like a fisherman fishing with rod and with reel.
It coaxes its catch,
Even wiggling the bait,
And knows very well there may be a long wait.

What it lures and draws close
It sucks into its mouth,
Where its teeth start their chewing
From north and from south.

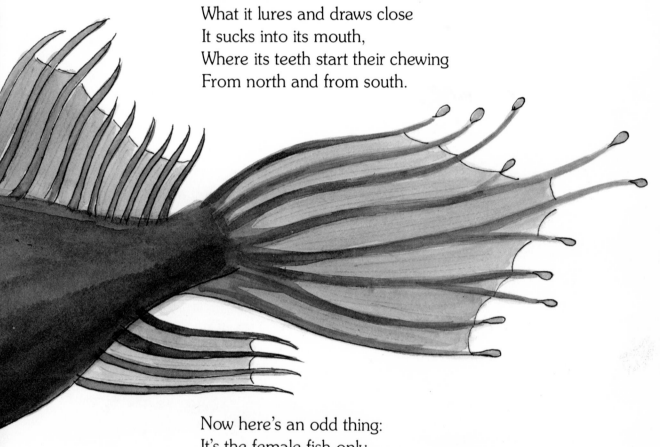

Now here's an odd thing:
It's the female fish only
That does all the fishing, yet never is lonely.
For its husband hangs onto its underside
And there very happily hitches a ride.
The female won't shed it
By wiggling and swishing.
It is, as you'd guess,
Far too busy with fishing.

JAPANESE SPIDER CRAB

Here's one you might find,
But it's doubtful you can,
For it's deep in the ocean surrounding Japan.
If it rarely is seen,
There's no reason to fret,
For it's hardly a creature you'd like
For a pet.

This Crab is the largest of crabs in the world
And looks even larger
When legs are unfurled.
Its body is big as a man's
But it looks
Much larger because of those legs
And those hooks,
Which are pincers for pinching and holding quite tight
Till the Crab's in position for taking a bite.

It has legs like a spider's
And yet is a crab,
And when it is crabby
It's quick on the grab.
Twelve feet between claws, it has quite a long reach,
And anything grabbed
Would let out a loud screech.
Unless, and the chances for this may be scant,
The thing that is grabbed is an innocent plant.

SEA LAMPREY

If you think a Sea Lamprey is big,
You are wrong,
Since as big as it gets is a mere two feet long.
Its twisting and turning is helped by the blubbery
Backbone it has.
It's conveniently rubbery.
No scales has this creature,
It's smooth as an eel.
You can prove it yourself
If a Lamprey you feel.
There's something else also
That's cause for hurrahs,
The Lamprey is lacking
Completely in jaws.

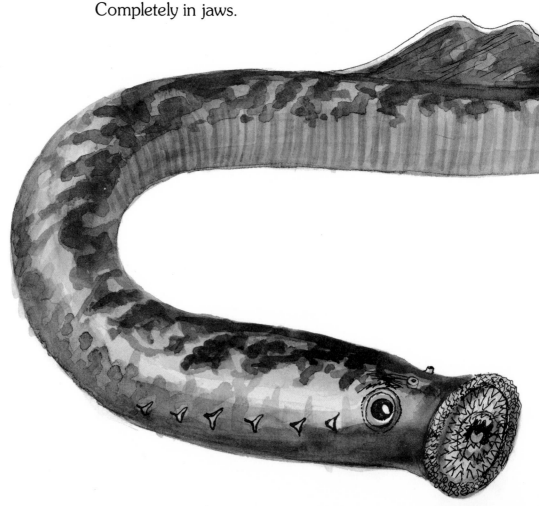

And yet, for all this, it's a fish causing fear
To fish that are larger, when Lamprey comes near.
And what is the reason?
In case you have wondered,
It has muscular lips
Lined with teeth—a whole hundred!
It cuts its way through
Scales and skin of a fish,
And then does a thing that no fish would wish.
In fact I can't think
Of a thing that is meaner:
It sucks out the blood
Like a strong vacuum cleaner!

Yes, the Lamprey likes blood
And it sucks some and swallows,
After which it sucks more,
Which the next swallow follows.
They call it the Lamprey, and that it may be,
But it's also like Dracula, deep in the sea.

GIANT SQUID

The Giant Squid
Is truly an oddy.
Eight arms and two tentacles stretch from its body.
Shaped like a cigar,
It's equipped at its rear
With a fin that is flat
And with which it can steer
And also dive down, as it does in the day,
Or come up when it's night,
In its up-and-down way.

The tentacles, fifty feet long, on their ends
Have claw-studded suckers,
Not mittens, my friends.
What the tentacles seize
In the Squid's arms they place,
And the arms, suckered too,
Lend a loving embrace.

The arms and the tentacles, north ones and south,
Are placed in a circle around the Squid's mouth,
So the Squid needs no napkin
But eats with its beak
Enough, it would seem, to last for a week.

The Squid doesn't swim with its arms, as you'd think,
It's as if jet propelled, thanks to storing a drink.
Yes, it takes water in
And then squirts water out,
Thus moving, torpedo-like, swiftly about.
And not only that,
But if danger it fears
It squirts inky stuff
Till its shape disappears.

With tentacles, arms,
And with suckers and beak,
A Squid once did something of which sailors speak.
It reached up and grabbed,
For a reason unknown,
A full-masted schooner and made it its own.
It pulled it below to the depths with its dive,
And only the captain was rescued alive.

So one thing be glad that you never once did:
Step up and shake hands with a strong Giant Squid.

LARGETOOTH SAWFISH

Well-named is the Sawfish one looks at with awe,
For its snout, long and flat,
Seems a double-edged saw.
To the Shark it's related, like mouse to a rat,
But no Shark has a nose that's the least bit like that.
If it lived on the land
And not in the sea,
The Sawfish could easily cut down a tree.

The one that's called Largetooth
Is twenty feet long,
Weighs a full thousand pounds
And is strong,
Very strong.
Yet it mostly is found in the western Atlantic
Up close to the coast, and more lazy than frantic.
It stirs with its saw
In the ooze of the bottom.
Any crabs and the like?
If so, it's soon got 'em.

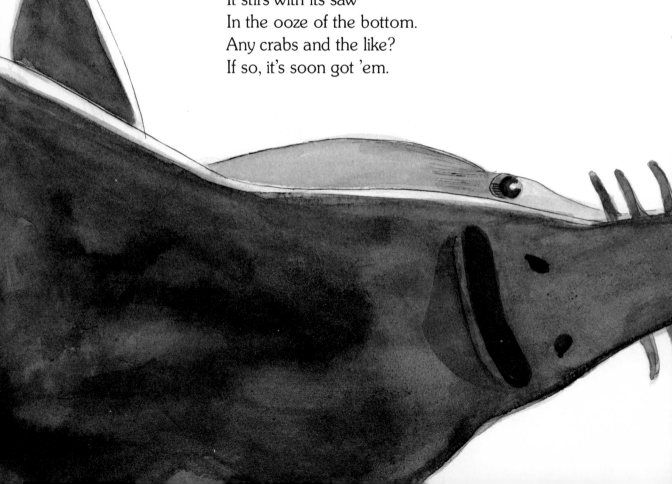

But sometimes when spying a school of nice fish
It slashes about with a swoop and a swish,
Then eats at its leisure,
Till hungering ceases,
What falls to the bottom,
Both large and small pieces.
Some tell of its sawing a swimmer in two,
Or maybe a boat,
And who knows if it's true?
It may not compare with a Shark as to jaw,
But on no other fish
Will you see such a saw.

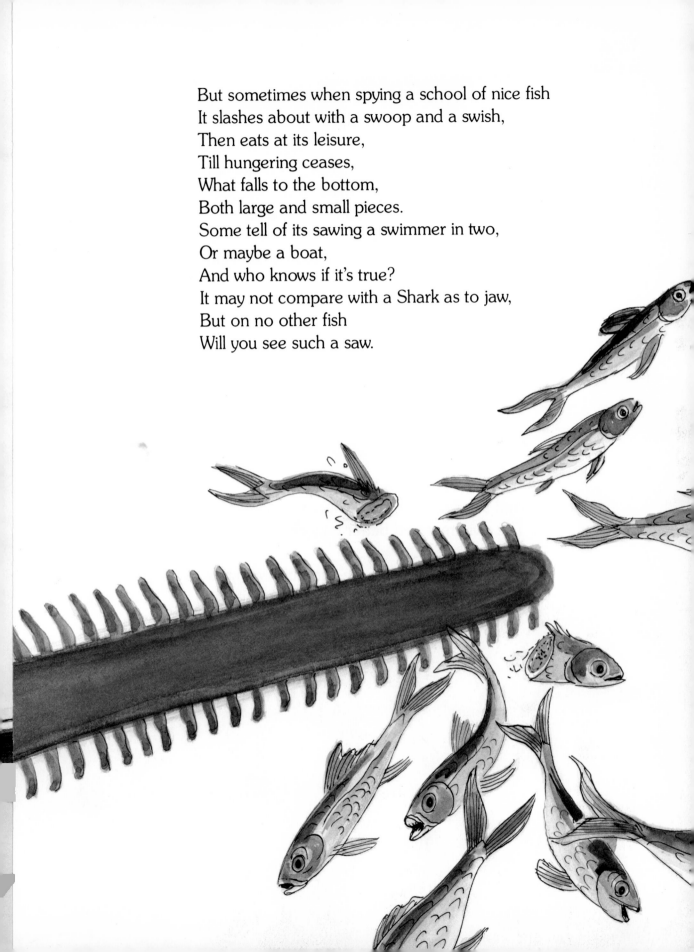

GREAT BARRACUDA

Some Barracudas are very small
And won't hurt you much
If they hurt you at all.
But you'd better beware of the one called Great.
That one's ten feet long
Or anyhow eight.

This Great Barracuda is swift and it's fierce,
With long, jutting jaws
And with teeth that will pierce.
It appears of a sudden
And bites a deep gash,
Then it goes as it came,
Disappears in a flash.

Built nicely for swimming,
It's long and it's lean,
But it's also a fish
That can be mighty mean.

It likes best the ocean in places most warm,
But also it's fond of a leg or an arm
Or anything else that is fresh, juicy meat,
And that includes fingers
And hands—
Yes, and feet.
Since it's fish it is after, you *might* feel secure,
But in water that's murky it's not always sure.
After all, it is known
As the "lion of the sea,"
And a name such as that is sufficient for me.

I suggest when you swim that you leave behind
Any bracelets or buckles
Of glittery kind.
And don't wear a bright-colored bathing suit
That might well attract
This scissor-toothed brute.

The chances are slight that it ever will hurt you,
But taking precautions I think is a virtue.

MORAY EEL

Many the kinds of eels there are,
As many, indeed, as the shark or the gar.
The Moray, however, puts up the most fight
And possesses the savagest,
Wickedest bite.

It's a bright-colored eel
With a leathery skin
So tough that a knife
Would be hard to get in.
And yet it is slippery, hard to hold onto,
Assuming, of course, anybody would want to.
It's a full foot around and some ten feet long,
With muscular jaws and with teeth large and strong.

It lives in warm waters,
Perhaps in a hole,
Or writhes between rocks,
Where it hides like a mole.
It hides just to rest,
Not because it is frightened—
Just to rest till for eating
Its stomach is lightened.

It tells by the smell, though some distance away,
Just where and just what
Is its edible prey.
Its favorite food, as an octopus knows,
Is an octopus, which to a hiding place goes
And hopes for the best
And quivers with fear
When the Moray comes sliding and slithering near.

But hiding won't help,
Since the Moray can stick
Its sharp-pointed head right inside
And quite quick.
It will nibble the octopus, nibble and pull,
Till the crevice is empty,
Its stomach is full.

The Moray won't hurt you, they say, unless you
Hurt it first, and I certainly
Hope this is true.

DRAGONET

The Dragonet is hardly a dragon.
A little fish, its size it can't brag on.
Its head is as flat
As if squashed by a stone,
As lowbrow a head as you've ever known.
And its eyes are both placed in an odd sort of place:
On the top of its head
(Or the top of its face).
If you faced its face, which I doubt you would do,
You couldn't be sure it was looking
At you.

But you can't tell by looks,
And this small, ugly thing
Has just what it needs
When it wishes to sting.

When the Dragonet is prepared to attack,
It thrashes its tail.
Like a fan,
Back in back,
And that swings a murderous spine to and fro,
And anything hit cries out, "Please,
Oh, no!"
For the Dragonet's spine is something quite serious
And leaves what is stung either dead or delirious.

The Dragonet male can turn colorful hues,
And a female's impressed
And can hardly refuse.
But unless you're a female Dragonet,
Get away just as far
As you're able to get.

OARFISH

A feature that's odd
And that gives it its name
Is its fins shaped like oars,
Which no others can claim.
Since they're thin and they're round
And have paddle-like ends,
Are they used, then, to row with? I'm sorry,
My friends.
In spite of their oar-shape, which causes much peering,
They're used not for rowing at all, but for steering.
The Oarfish, whatever its personal wishes,
Swims by wriggling like eels
And like most other fishes.

Another strange thing is the shape of its head.
It's shaped like a horse's,
Or so I have read.
And up from its head, that is otherwise plain,
Is what is described as a flaming red mane!
In fact a red fringe will be found,
Never fail,
On its back from its head to the tip of its tail.

The Oarfish is thirty feet long,
Which of course
Is longer by far than the length of a horse,

And also it lives where no horses would be:
Up and down in the Mediterranean Sea,
As well as Pacific and most other oceans,
There weaving its way
With its sinuous motions.

Some have seen it move snake-like
Upon the sea's surface,
With horse-head and mane,
And it made them quite nervous.
If it isn't a monster to flee from in fright,
It *looks* like some sort of a monster
All right.

Far down in the deepest depths of the sea
Food is hard to get,

BLACK
SWALLOWER
As hard as can be.
So a fish must accomplish remarkable feats
To get, and get down, all the goodies it eats.

Consider the Swallower
Known as Black,
The color that covers it,
Front to back.
Like the fish that's called Gulper, it can't get enough,
And eats what is tender
And eats what is tough.
In fact like one cannibal eating another
A Swallower swallows its Swallower brother.

But strangest of all, and the greatest surprise,
Is its swallowing something that's twice its own size!

They say that off Spain,
Where they caught one
One day,
It does this in quite a remarkable way:
When it opens its mouth,
Its jaw up on top
Swings wide on a hinge
While the lower's let drop.

Its backbone it bends,
It moves many a part,
Including its gills and even
Its heart.
Then down goes the mixture of bones, blood, and jelly
To stow in its stomach or swollen-out belly.

Imagine a wolf gobbling down a deer whole
And you'll see what must be
The Black Swallower's goal!

SABER-TOOTHED VIPERFISH

Now here is a fish that though smallish is scary,
At most a foot long,
But you'd better be wary.
And the reason is this:
It mostly is head,
To make room for a mouth
That its prey must dread,
A mouth it keeps open because, I suppose,
It's simply too full of big teeth to close.

Another reason could be that inside
This mouth that is always kept open wide
There are lights that shine
Very bright
And clear,
Telling customers "OPEN"
or "ENTER HERE."

On the Viper's sides there are other lights,
Two rows that a curious passerby sights,
And then coming closer
To have a good look,
Is grabbed by a fang
That acts like a hook.

As if *that's* not enough, on its chin,
My friend,
There's a whisker it trails
With a light on the end.
With all of these things and with jaws so strong,
The Viper should never be hungry long.

DEADLY
STONEFISH

The Stonefish is one
(And one fewer I'd wish)
Of the family known as the Scorpionfish.
They all are alike in at least one thing:
They've a horribly painful
And deadly
Sting.

But the Stonefish, to give you my honest opinion,
Is the ugliest creature in sea world's dominion.
Its skin is all covered,
Yes, all of its parts,
With rough-looking, hard-looking,
Terrible warts.

In the Indo-Pacific, the place where it's known,
It seems just a chunk of old coral
Or stone.
And so it can hide in the rocky debris
And never be noticed by you or by me.

But if it is touched, it lets fly
With its spines
A sting that's far worse than a porcupine's.
Its spines are so strong
That it's said they'll go through
The thick rubber sole of a tennis shoe.
Thus the spine carries deep
Such a powerful shot
Of poison that what
Was alive now is not.

Ugly-looking the Stonefish,
It's easy to tell,
And believe me, it's ugly
Acting as well.

ZEBRAFISH

The Zebrafish is called Lionfish—
Call it Turkeyfish, Tigerfish,
Too, if you wish.
It's a scorpionfish,
As the experts have stated,
And thus to the horrible Stonefish related.

But the Stonefish is ugly,
As any can see,
While the Zebra is pretty
As pretty can be.
It's a colorful fish, mostly some sort of red,
But also you'll find,
From its tail to its head,

That it's streaked with black marks,
And to make it more bright
Has patches, like snow,
Here and there,
That are white.

When it spreads out its fins and its spines that abound,
It struts like a turkey or peacock around.
You might well be tempted
To pet it and say,
"Come here, little Zebra,
Come here and let's play."
But the Zebrafish uses its looks much like bait.
One touch of a spine, and your love turns to hate.
And no venom they lack.
When the Zebrafish fights,
It is stubborn and tough.
It spears with its spines
Full of poisonous stuff.

So look for the Zebra in water that's warm,
And don't be misled by its colorful charm.

MARINE IGUANA

The Galápagos Islands, off Ecuador's shore,
Is the place where you'd find
What you're looking for:
The Marine Iguana, a creature that looks
Like Tyrannosaurus in dinosaur books.
It's smaller, of course, maybe four feet long;
If you said, "A huge lizard,"
You'd not be far wrong.

Ferocious it seems,
With its rough,
Bony face,
And with claws on its feet
And long tail swung with grace,
And with sharp, thorn-like thistles
On head and on back,
And a color unpleasant,
A dull, dirty black.

In the water, it swims through the terrible tides
In a serpent-like way,
Legs tucked in by its sides.
When it's not in the water, it basks in the sun,
And on rocks nice and warm
Seems to cook
Till it's done.
Whether swimming or sunning,
Iguana has company,
And never is noticed
To claw, bite, or bump any.
Though wicked it looks
With those teeth and those claws,
It eats only seaweed,
Not flesh. (Some applause?)

But when it's annoyed and upset
Or feels hurt,
From its nose spurts a vapor or some sort of squirt.
Then it looks like a dragon,
A dragon indeed,
And you might flee with fright,
But there's really no need.

GIANT GROUPER

One fish divers fear
Off Australia, we're told,
Is the Grouper. It's big, very big,
And it's bold.
Some two thousand pounds
(Or a ton)
It may weigh,
And its mouth when wide open
Seems big as a bay.

It's that mouth divers fear, even divers not thin,
For the Grouper can suck a whole diver
Right in.
Yes, the Grouper sucks food in
(One watches with awe)
As if sipping a soda,
Sucked up through a straw.

And here's an odd fact
That you maybe don't know:
The deeper live Groupers
The bigger they grow.

Some Groupers have spotty designs,
Some are plain.
They've fluttery fins, but their mouth is a pain.
A Grouper is not to be toyed with or followed
Unless you would think it great fun
To be swallowed.

PORTUGUESE
MAN-OF-WAR

No warship, the Portuguese Man-of-War,
But those it has stung think it worse by far.
Some it stings it makes tingly,
Some it stings it makes ill,
And some with its terrible sting
It can kill.

The Man-of-War fools you,
For what you can see
Afloat on the surface
Looks nice as can be.
The size of a football, it's bluish
I think,
With a crest, like a sail, with a border of pink.
It moves with the wind, and seems up to no wrong,
Like a floating balloon, it just bobbles along.
It may be alone or it may be with dozens
Of friends and of relatives,
Brothers and cousins.
In fact it itself is not one thing at all,
But a colony made up of parts very small.

But what you don't see, much like strands of blue string,
Are tentacles dangling beneath the top thing.
Pulled up, they can look
Like mere stubs, but let dangle
They're a hundred feet long, yet they don't seem to tangle.
The tentacles must have long muscles,
Unbeatable
For pulling up food that it hopes will be eatable.

To swimmers, however,
The number one thing
Is Man-of-War's terribly
Poisonous sting.

If a tentacle's touched,
Even brushed without vigor,
Off go pistol-like cells
That seem fired by a trigger.
Instead of a bullet, though, venom comes out,
As strong as a cobra's,
At least just about.
This fills up a hole a sharp spine has prepared,
And what happens then I won't tell
If you're scared.
It's well to be careful and always suspicious
Of things that look pretty
And *are* pretty—vicious.

COELACANTH
(SEE-la-canth)

The Loch Ness Monster may never be found,
But other odd creatures, it seems, are around.
Though many may tell us such things don't exist,
Some really are living
That searchers have missed.

Consider the Coelacanth,
Called that in Greek
And hard to pronounce
If Greek you don't speak.
As early as dinosaurs,
Even before,
It lived, but was thought
Not to live any more.
Coelacanth bones were dug up,
Old and docile,
And then put together
And labeled "A Fossil."
But a few years ago,
Off east Africa's coast,
They found a live Coelacanth,
Not just a ghost!
Since then, more and more have been found, I've been told,
And many are young, though they look very old.

The Coelacanth's bluish and five feet in length.
For something thought dead, it has plenty of strength.
What's strangest about it are tassel-like fins
And a tail with a tail,
(Where one ends, one begins.)
Those tassel-like fins may have helped it to crawl
On the sea-bottom once,
And not stumble and fall.

Another strange thing
About which I have read:
It's double-jointed inside of its head!
Since Coelacanth's bony,
It doubtless needs joints
At a number of rather unusual points.

The lesson we've learned when we've Coelacanth scanned
Is that species last longer in sea than on land,
And a fish that we thought
Had died out long ago,
If asked, "Are you dead?"
Might wave fins and say, "No!"

These are some of the creatures,
By no means all,
Some of them large
And a few of them small,
That live in the depths of the miles-deep sea
Or rise to the surface on some sort of spree.

If they're monsters, it isn't because of their size,
But because of the wicked look in their eyes,
And the way they can bite
Or the way they can sting
Or the way they can grab
Any living thing,
Especially anything smaller or weaker
Or slower
Or softer
Or careless
Or meeker.

What they do, how they do it,
You've seen in these pages.
It's what they've been doing for ages and ages.
They play, they have young,
But I'd like to repeat
That mostly their business, each day,
Is to eat.

Let's leave them there eating whatever they wish;
Be glad it's not you that's their food,
But a fish.
And don't go exploring and poking around,
Or you may disappear
And never be found.